Mathias Roth

The Physiological Effects of Artificial Sleep

With Some Notes on the Treatment by Suggestion

Mathias Roth

The Physiological Effects of Artificial Sleep
With Some Notes on the Treatment by Suggestion

ISBN/EAN: 9783337417277

Printed in Europe, USA, Canada, Australia, Japan

Cover: Foto ©berggeist007 / pixelio.de

More available books at **www.hansebooks.com**

THE PHYSIOLOGICAL EFFECTS

OF

ARTIFICIAL SLEEP

WITH SOME NOTES ON THE

TREATMENT BY SUGGESTION

BY

Dr. MATHIAS ROTH.

London:

BAILLIERE, TINDALL & COX,
20, KING WILLIAM STREET,
STRAND, W.C.

1887.

INTRODUCTION.

ABOUT 52 years ago, while a medical student in Vienna, I caught cold and had the misfortune of being treated according to the fashion of the day by bleeding, and besides the venesection as many leeches as could be placed on my arm were applied. My two very kind and xcellent teachers of anatomy, Professor Berres and his assistant, Dr. Hyrtl, who had afterwards a world-wide fame as a professor of anatomy, were my medical attendants. Their heroic treatment had a baneful effect on my constitution for years; thus I was induced to study various other, at that time, prevalent modes of treatment, amongst these animal magnetism. In the winter of 1839-40 I was present at the magnetic demonstration of Baron Dupotet at the Athenæum in Paris. A few years later Baron Reichenbach published his first experiments on odism. At the same time I had an idiopathic somnambulist patient under treatment who for some time after the attack suffered from periodical catalepsy.

On this patient I had occasion to verify many of the experiments made by Reichenbach. My interest in the study of animal magnetism was kept up by reading several works on the subject, and especially that of Dr. Ennemoser. In my long practice I have often had opportunity of relieving many patients from their pains and nervous attacks by the apposition of my hands or by passes.

In the London Mesmeric hospital established by Drs. Elliotson and Ashburner, I had opportunities of personally convincing myself of the curative effects of animal magnetism.

The *Zoist*, edited by Elliotson, gave an account of many remarkable cures, and of Dr. Esdaile's successful operations on patients in artificial sleep. With Dr. Burcq I have made many observations on the application of various metals in anæsthetic patients. Being engaged for many years in my Kinesi-therapeutic speciality, I could not follow the progress of animal magnetism in the form of Braidism and Hypnotism or the Treatment by suggestion.

Only last year I have read the works of Liébault, Bernheim, Bottey, Azam, *The Revue d'hypnotisme* by Barillon, and many other works of which a very complete catalogue is found in Dr. Ladame's book *La Neurose Hypnotique*, which list is continued to 1881, but since that time many additional and interesting works have been published in Germany and France.

Although I had prepared a copious compilation, I am publishing only this short pamphlet with the sole purpose of rousing the interests of my younger colleagues in a branch of the healing art which has made lately very great progress, and which undoubtedly will contribute to the cure of many diseases previously unsuccessfully treated by other methods.

M. ROTH.

June, 1887.

Wimpole Street, London.

THE PHYSIOLOGICAL EFFECTS OF ARTIFICIAL SLEEP WITH SOME NOTES ON THE TREAT-MENT BY SUGGESTION.

By M. Roth, M.D.

Artificial sleep was induced by Mesmer and his followers, who believed that by "passes" or movements of the hands from above downwards, a magnetic fluid passed from the body of the mesmeriser to that of the patient. The idea of such magnetic fluid passing was first denied by Faria, a Portuguese Indian priest, and in 1843 by James Braid, a Manchester surgeon, who showed that fixing the eyes on a bright object held before and a little above them induced hypnotism. These experiments have since been widely studied and of late years chiefly by Charcot and his pupils; one of whom, Dr. P. Richer, has published the results of Charcot's enquiries. Professor Heidenhein of Breslau, who witnessed the experiments of Hansen, has also carefully investigated the subject, and added fresh observations. Lately, Dr. Bremaud, a French naval surgeon, has, in addition to the stages of hypnotism, lethargy, catalepsy and somnambulism, described one he has observed, in men only, as "fascination." Liébault, who made use of suggestions as a therapeutic agent twenty-five years ago, says that 90 per cent. of healthy persons are susceptible to this influence. Dr. Bottey found 30 per cent. of healthy women to be susceptible. Ochorowitz observed that those who were easily affected by a mineral magnet were also susceptible to hypnotism. This has not been verified by other observers.

There are various methods of inducing hypnotism. The eyes are fixed on some bright object in front and a little above the level of the eyes. Fatigue induces a slight convergent squint. The eyelids are closed by the operator making slight pressure upon them, and by placing the palm of the hand upon the forehead the lids continue closed. Silence, the full consent and absolute tranquility of the patient are indispensable conditions for success. If, on a first experiment, hypnotism is not induced in from three to twenty minutes, the experiment must be relinquished and repeated on a future opportunity. When once it has been induced it will be brought on rapidly in future experiments.

Professor Bernheim lays great stress on the importance of gaining the full confidence of the patient, convincing him of the entire harmlessness of the process, and giving some explanation of what will occur before commencing an experiment. While the patient is gazing fixedly at him he endeavours to concentrate all his attention on the idea that he is going to sleep. " I then say," he continues, " your eyelids are closing ; you cannot open them. Your arms and legs seem heavy. You cannot smell anything. Your hands are immovable. You cannot see. You are going to sleep. I then add, in a somewhat imperative manner : Sleep ! Often that one word is sufficient ; the eyelids close, the patient is asleep." In cases where patients are not very tractable, when they do not think of going to sleep, and the eyelids do not close, the Professor, after looking fixedly at his patient for two or three minutes, closes the eyelids, and keeps them closed, and then addresses the patient, concluding with the command : " Sleep ! " This, he says, " I call sleep by suggestion. It is the image of sleep that I insinuate on the brain."

Children, when once they understand what is said to them, are easily hypnotised, as indeed are some adults, simply closing the eyelids, keeping them closed, and uttering the command " Sleep " being sufficient.

In obstinate cases, Professor Bernheim keeps the eyes closed, insists strongly on silence and immovability, and constantly assures the patients that they feel numb, listless, that the extremities are stiff, that they are asleep, &c. After frequent repetition of these assertions, the eyelids remaining closed, the fingers are withdrawn, he lifts the arms and they remain extended ; this is cataleptic sleep.

With some, success is more completely obtained by persuasion ; others must be addressed sharply in order to prevent resistance. After one or more sittings, patients are generally easily hypnotised. In many it is enough to place the fingers before the eyes and say " Sleep." When the tendency to be hypnotised is strongly developed, the suggestion can be conveyed by correspondence, it being sufficient simply to assure such

that they will sleep as soon as they have read the letter.
M. Liègois has succeeded in conveying the suggestion to
sleep by telephone, all that is necessary being to convey
the suggestion to the mind by some means or other.
Hypnotism presents itself under three stages. 1st.
Lethargy, a condition of deep drowsiness, with loss of the
use of the senses. 2nd. *Catalepsy*, a state of automatic
muscular rigidity. 3rd. *Somnambulism*, a loss of conscious cerebral activity, with exaggerated certain cerebral
functions. 4th. *Fascination*, observed by Dr. Bremaud.
Twenty-five per cent. of healthy hypnotised persons pass
through all these stages. In a larger number only
lethargy and somnambulism are observed. In a third
proportion only somnambulism occurs.

LETHARGY. This state can be produced by either of
the methods already described: After catalepsy by
closing the eyelids or blowing gently on the eyes. After
somnambulism by friction with one or two fingers on
the top of the head, by closing the eyelids and pressing
gently on the eyeballs, or by gently blowing on the
eyelids.

During lethargy, a person is in a complete state of
languor ; the head is inclined to one side ; the extremities
are relaxed ; eyelids tremble ; eyeballs turned upwards ;
the patient snores ; the skin is insensible to any stimuli ;
anæsthesia is constant ; the special senses are perfectly
dormant ; some have observed acceleration of the heart's
action ; Bottey did not, but a slightly flushed face and
epistaxis ; pressure roughly applied to the muscles, as by
kneading, causes their immediate contraction. This
property corresponds to Charcot's neuro-muscular hyper-
excitability. Thus when the brachial muscles are kneaded,
the forearm becomes bent and remains so. To release
the forearm it is necessary to apply similar pressure to
the antagonistic muscles. Contraction thus excited can
be transferred by placing a mineral magnet near the
limb of the opposite side. Brissot and Richer have proved
that no such contractions will take place in a limb
kneaded while in a state of bloodlessness through the
application of Esmarch's bandage, but they occur directly
the bandage is removed. Charcot and Richer have
repeated this experiment, and have shown that this dormant mechanical excitement of a muscle can be transferred. Having produced anæmia in the right arm by
Esmarch's bandage during the lethargic state, and excited

the cubital nerve without producing any contraction, they placed a magnet near the left arm, and at once the latent contraction of the right arm appeared in the left. In persons who are readily placed in a cataleptic state superficial excitement of the skin will produce similar contractions. In some rare cases of artificial catalepsy contraction of one side of the body can be induced by blowing strongly into one ear or tickling one nostril. The application of a galvanic current to one half of the head produces certain movements in the limb of the opposite side (Erb).

By gently raising the eyelids and blowing on the eyeballs the state of lethargy comes to an end. The transition from hypnotism to the waking state is sudden, sometimes there is a slight twitching of the eyelids for a few seconds and the subject rubs his eyes before he is quite conscious that he is awake. No mental anxiety or heaviness of the head is felt on awakening.

CATALEPSY occurs early in hypnotism, and passes into lethargy when the hypnotising process has been continued too long. When induced by fixing the eyes steadily on any object, the moment when the eye appears fixed and the conjunctiva injected, the object must be removed. If catalepsy has been induced by pressure on the head, the hand must be removed the instant catalepsy is developed. If the pressure is continued lethargy takes place.

A more or less intense sound or light, such as lightning, induces catalepsy. In highly sensitive persons, who have been the subjects of repeated experiments, a very slight excitement as, a light, the noise of two glass tumblers striking each other, or the sudden fall of some object will induce catalepsy. If these causes were more intense lethargy would ensue, showing that the two stages are produced by simply varying the intensity of the exciting cause. Lethargy, again, can be converted into catalepsy by opening one or both eyes of the subject in a well-lighted place.

In catalepsy the subject is suddenly fixed in one position; the eyes are fixed, pupils dilated, cornea insensible, eyelids open and twitching, no reflex action can be excited in them; the skin is insensible to pain; the special senses are to a certain extent impressionable, and conversation is possible; the limbs, though supple, will retain, for a longer or shorter period, positions contrary to the laws of gravity; the muscular sense is so far exaggerated that its contractile power increases in proportion to the resistance it has to overcome. Thus the body remains rigid when the back of the head is placed on one chair and the feet on another. Contractions induced during lethargy will continue during catalepsy. Contractions are excited by superficial rubbing of the skin. The so-called passes of mesmerism act in a similar way, viz., by superficially exciting the skin.

Catalepsy can also be induced on one side of the body while the other is in a state of lethargy by opening one eye of the subject. The opening of the right eye induces catalepsy of the left side of the body. On the cataleptic side the limbs are rigid, while on the lethargic they remain relaxed.

The following experiment illustrates this point. Place some crochet work in the hands of a person in an artificial cataleptic state, and the work goes on as if she were awake. Close the left eye. Now the corresponding arm begins to move slowly, and finally sinks down inertly, while the right arm continues the necessary movements.

The magnet placed a few centimetres from a lethargic arm, the other being cataleptic, renders the lethargic arm cataleptic, and that which was cataleptic lethargic.

Catalepsy ceases as soon as the eyes are strongly blown upon, and the subject on awakening has no recollection of what has occurred during sleep. When, as in some cases, blowing on the eye fails to arouse, the catalepsy must be converted into lethargy or into somnambulism. Catalepsy can be changed into somnambulism by blowing gently on the eyes, closing the eyelids, or by pressure on the vertex. When catalepsy is primary these methods generally induce lethargy, when it is secondary somnambulism with closed eyes, or, more rarely, somnambulism with open eyes.

Suggestions during catalepsy.

By a suggestion during catalepsy is understood an impression made on the brain of an hypnotised subject by commands, assertions or declarations. Suggestions made during the waking state are the result of an idea taken up by the subject or proposed by another person

During the cataleptic state the attitude in which the subject is placed is reflected on his countenance ; it appears to suggest to the brain the sentiment it represents. If the hand is placed near the mouth, as though a kiss were to be despatched, the subject smiles. If the body is placed in an attitude of fear, the same feeling is expressed by the face. In some rare cases the expression the face is made to assume is reflected in the attitude of the body. If the zygomatic muscles are excited by electricity, the person seems to laugh and the attitude of the body at once corresponds. In Duchenne's work, in the plates representing the various feelings excited by electricity, the position of the body will be observed to correspond to the expression of the face.

During catalepsy, if an object is kept swinging before the eyes they lose their fixity and follow its movements. A floating object will, in this way, convey the illusion of a bird flying and excite a smile, and the hypnotised will sometimes try to catch it. If creeping movements are made disgust is felt and expressed by the countenance, as though a serpent were moving before the subject. As soon as the external influence ceases, he relapses into the rigid condition and the eyes are again fixed.

When the limbs of the cataleptic are set moving in a rhythmical manner, as when " beating time," as soon as the assistance rendered in making them is removed the motions continue automatically until a hand is rapidly moved before the eyes, when they cease. Whatever the operator does is automatically imitated by the hypnotised cataleptic, even to the extent of repeating the words of a foreign language in the way he hears it, and with much precision, if the language is known to him.

Memory also acts automatically. Thus, if an object is placed in the cataleptic's hand, he will make such use of it as he is accustomed to do. If, for example, a tumbler is given to him he will appear to drink. A spontaneous suggestion is caused in the brain by the appearance of a familiar object.

The senses being preserved, the experimenter is able to put himself in communication with the subject by speaking to him in a more or less loud voice, and will receive answers. Illusions, suggestions of various hallucinations, paralysis, contractions, &c., can then be excited. Some suggestions may be made which will be operative after the hypnotic state has passed off.

Somnambulism is of two kinds, one when the eyes are closed and the other when they are open. The means of producing lethargy primarily can also cause somnambulism with closed eyes. In cataleptic subjects somnambulism is more frequently observed as a primary state than lethargy. Somnambulism with closed eyes is produced after catalepsy by closing the eyelids, by gently blowing on the eyeballs, or by rubbing the head, and it follows more frequently if the catalepsy is secondary. It follows after lethargy by rubbing the vertex, and sometimes by blowing strongly on the eyes. It very much resembles lethargy.

During *somnambulism with closed eyes* the subject seems inert, is insensible to external objects, the head is inclined to one shoulder, the limbs are not so completely relaxed as in lethargy; the inertia is more apparent than real, as directly the somnambulist is questioned he answers, and will do whatever he is told; the eyelids quiver more or less; anæsthesia of the skin is generally complete, though it will vary in degree in the same subject at different *séances*. In some instances hyperæsthesia has been noticed. Azam reports a case where a lady felt a certain amount of heat when the naked hand was placed at a distance of 40 centimetres from the back. More or less hyperæsthesia is generally observable in sight and hearing. Though there be but a scarcely perceptible opening between the eyelids, an object placed before the eyes is recognised. The acuteness of vision in somnambulism has been tested by preparing a number of little squares of white paper, each containing a scarcely visible mark, and still it has been detected (Berger, of Breslau). Bottey quotes a case of hysterical blindness, in which sight was completely recovered during a state of spontaneous somnambulism.

Equally acute is the sense of hearing, of which many illustrations have been given by Bremaud, Berger and Ladame. At the same time, it is quite easy to produce false impressions on the sight and hearing by suggestion. The dynamometer indicates a very considerable increase of force in the somnambulist.

Intellectual hyper-excitation is also remarked during somnambulism. Bremaud mentions the case of a pupil at the Lyceum who, during a state of somnambulism, solved quickly and accurately a difficult trigonometrical problem, which gave him much trouble when in his normal state. Many similar cases are recorded by different authors. Neuro-muscular hyper-excitability is constant. Order a person to throw a stone, the arm is raised and remains fixed in the attitude of doing so. Superficial excitation of the skin always produces contractions in persons brought into a state of catalepsy, when hypnotised. Of the events immediately preceding hypnotism all remembrance is generally, but not always, destroyed, while the memory retains whatever occurs during hypnotism. Braid mentions the case of a woman who, during somnambulism, recited passages of Hebrew and other languages, of which she was quite ignorant when awake. On enquiry it transpired that when a young girl she had heard a clergyman, in whose house she resided, read those passages aloud. Richer quotes the case of a somnambulist who sang an air from *L'Africaine* which she had heard only once and was unable to sing when awake. Dr. Pope, when a student at the University of Edinburgh, heard a man sing a Gaelic song while lying on the operating table, under the influence of chloroform. After the operation, a Highland student went to him and addressed him in Gaelic. He could not understand a word, and said that he had entirely forgotten the language, never having spoken it for forty years. Bottey read some paragraphs from a book two or three times to a person and afterwards hypnotised her. He then told her to repeat what he had read, and she did so almost verbatim, but when awake could not recall a word of either. Many similar experiments are on record.

This excitation of the memory occurs in various pathological and physiological states of the brain. Drowning persons, when death seemed imminent, have seen as it were the actions of their whole lives pass before them. *Opium, haschish, chloroform* cause a similar hyper-activity of the brain. I attended a lady in a state of spontaneous somnambulism, who, at a later stage, when the attacks of catalepsy and somnambulism returned periodically, used to give an account of many of the events occurring when she was four years of age, of which she had no recollection when awake.

During the state of somnambulism, all suggestions made to the subject will be acted upon. He is incapable of resisting them. Sometimes they exhibit an unnatural degree of strength, and it becomes necessary to rub the

vertex in order to change his somnambulism into
lethargy. All power of spontaneous action or exercise of
the will is destroyed; if left to himself he falls into a
state of inertia until aroused by being spoken to.

Somnambulism with the eyes closed ceases either by
the induction of the waking state or by its change into
the hypnotic state. Awakening occurs spontaneously,
but this is not advisable. Bottey saw one person awake
after a sleep of seven and another of twenty-two hours.
In some very rare cases, awakening takes place after
the upper eyelids have been raised; but a suggestion or
blowing on the eyelids is the most frequent mode of
awakening. Such blowing must be gentle and continued
for some time, or it simply changes the condition into
one of lethargy or of somnambulism with the eyes open.

Somnambulism with the eyes open is never a primary
state. Most frequently it is secondary to somnambulism
with the eyes closed, the mere raising of the eyelids
being sufficient to produce it. It is not a little remarkable that the somnambulist is unable, even by the
influence of a suggestion, to open his eyes himself.
During catalepsy rubbing the vertex will cause somnambulism with the eyes open.

While in this condition, the eyes are widely open, the
eyelids throb, the eyeballs are not fixed as in catalepsy,
but have a haggard appearance. In some persons the
countenance is unchanged. Anæsthesia of the skin is
constant. The senses are hyper-æsthetic. Neuro-
muscular hyper-excitability is always present. Contractions also take place from irritation of the skin in
catalepsiable persons. During this state there is no
knowledge of locality, but great illusions as to place,
one seems to think himself in a well lighted room,
another in a forest, another on the banks of a river and
so on. There is a certain unconscious activity, and
incessant desire to move; if left to himself he rises,
walks, uses his senses and sometimes offers a degree of
resistance to the will of the operator. Suggestions of
any kind can be made effectively to those who are
amenable to hallucinations and illusions.

The following illustration is given by Bottey :—

M. G. and A. L. were hypnotised and left to themselves when in the stage of somnambulism with the eyes
open. M. G. at once imagined herself in the parterre of
a garden, and exclaimed, "Oh! what pretty flowers."

She stooped as if to gather them. A. L. walked about believing that she was in the mud, and raised the skirt of her dress to avoid its being soiled by it. After a while the two walked together to the bank of a rivulet, when one imagined she saw a snake and wanted to run from it, while the other stepped forward and, although trembling, acted as though trying to kill the reptile. Finally they sat down on the bank and one of them took a foot-bath. At that moment one of the persons who assisted at this little scene stepped forward to speak to them. A. L. immediately had the illusion that a gend'arme was approaching, who told her that they were trespassing, and threatened them with a *procés verbal.* Quite alarmed, she became pale, and both ran away and tried to hide in the corner of the room. Dr. Bottey now interfered, and suggested that he was the proprietor of the estate, and assuring them that they had no cause for fear invited them to take some refreshment; having taken numerous imaginary liquors they laid down in a hammock and fell into a lethargic state, from which they were awakened by blowing in the eyes.

Somnambulism with the eyes open ceases by returning to the waking state on the same conditions as that with the eyes closed. Closing the eyelids converts it into somnambulism with eyes closed. Prolonged closure of the eyelids, with slight pressure, produces lethargy, as also in some instances does intense blowing upon the eyes or rubbing the vertex.

Suggestions made by any person present at a *séance* are effectual. The somnambulist is perfectly automatic, and acts entirely in accordance with the various excitations to which his brain is exposed. Any object can be made the cause of an illusion. Tell a person that those around him are trees, and he will make climbing movements; tell him that a small piece of stick is a cigarette, and he will want to smoke; tell him that some ammonia you offer to him is a perfume, and, notwithstanding the hyperæsthesia of the senses, this even will for the moment be destroyed by the force of the suggestion. So also with the sense of taste. Sugar will be bitter or a bitter substance sweet, according to the suggestion made. Different hallucinations can be produced on each side of the body. One eye will see a flower and the other a bird; one nostril will smell a rose, while the other receives the odour of asafœtida. The two sides of the brain thus receiving at the same time absolutely opposite impressions, according to the suggestions made. By placing a prism before one eye, or pressing the external part of the eyeball, the double illusion vanishes. Blindness, dumbness, deafness can be produced in the hypnotised person by assuring him that he can neither see, speak, nor hear. Similarly the sensation of the skin can be altered ; the muscles of the extremities can be influenced in any direction. Without speaking, a limb set to certain motions will continue in them automatically.

Suggestion can be varied in every possible way, and will give rise to corresponding agreeable or disagreeable sensations and attitudes. Professor Heidenheim, of Breslau, gives some very striking illustrations of the power of suggestions. All who have been present at the public performances of the so-called electro-biologists and mesmerists will remember that the somnambulists were unable at the given suggestion to pronounce their own names, or certain letters, or to do the slightest things for themselves as long as the corresponding suggestion was not interrupted by another, or until they were awakened. On the suggestion to awaken being given in a tone of command, the subject rubs his eyes, blinks for a few seconds, and looks around in surprise.

If, during somnambulism, an order is given to do a certain thing at a date some time after awakening, it will be done, and that without the subject knowing or understanding why he is impelled to do it. Hallucination can be suggested that will come into operation after a longer or shorter period of time. Bottey suggested to a person that he should in a week's time ask him for the flower he had in his button-hole, and this was punctually done— the interval being a week.

This manifestation of the suggestion occurs either from its having been dormant in the brain cells and awakened at the fixed time, or it operates as a " possession " tormenting the person until the moment arrives when its fulfilment is due.

It sometimes happens that when a suggestion of a character naturally repulsive to the hypnotised subject is made great reluctance to act upon it is shown. A young nurse during the hypnotic state was ordered to take a fancy to the attendant of a dissecting room. When she awoke she began to cry and refused to give the cause of her grief. On being very much pressed she said that she was, by some irresistible impulse, compelled to like the attendant although she felt that he was both dirty and disgusting. She was then thrown into the hypnotic state, and the suggestion negatived in order to counteract this abnormal impulse.

Fascination is a phase of hypnotism described by Dr. Bremaud as having been observed by him in men. It was produced by fixing the eyes on a brilliant point of average intensity. During this stage the temperature and pulse rise; the pupils are dilated; analgesia is present; muscles rubbed by the operator contract violently; there is an irresistible impulse to follow the brilliant point on which the eyes are fixed; the will is paralysed; illusions and hallucinations arise; and an instructive imitation of movements, gestures, attitudes, physiognomy and words is developed.

Dr. Bremaud gives an illustration of this stage, of which the following is an abstract:—

A. Z., æt 23, is a dark, sanguine, vigorous man. I looked at him, and directed him to look at me with as fixed an eye as he could. The effect was rapid. His face was congested, eyes widely opened, pupils dilated, pulse 120. When I move back, Mr. Z. follows me. In walking the head projects forward, the shoulders are raised, the arms hang down. In running the arms remain immovable. All expression disappears from his countenance, not a muscle of the face moves, not a word is uttered. He seems unable to desist from staring at my eyes. Question him, he does not reply; insult him, not a muscle of the face moves; strike him, and he feels nothing; neither pinching nor tickling are felt. He is nevertheless perfectly conscious, and on returning to the waking state will recount all that has been said and done. Blowing on the eyes at once removes the fascination, and the subject immediately resumes his natural appearance. In another case the subject imitated everything done by the operator, laughing, jumping, grimacing, or speaking. When the influence was removed he did not remember anything that had passed. Recollection of what has been said and done is removed by the prolongation of an experiment.

Fascination is changed into catalepsy by directing the fascinated one to look steadily at a brilliant light, such as a lighted gaselier. The pupil remains dilated, but the face becomes pale and bloodless, and the limbs, though relaxed, remain in any position in which they are placed.

Bremaud believes that his failure to produce this stage in women is due to their great nervous impressionability rendering them unable to fix their eyes sufficiently steadily in the beginning, from which they fall into catalepsy.*

I now propose to consider the application of artificial sleep as a therapeutic measure.

Before I was acquainted with the progress recently made in this subject, I had frequently relieved pain, spasms, hysterical and other nervous attacks by making longitudinal passes in a centrifugal direction, and by placing the hand or a part of it on the painful spot. That hysterical attacks can be checked by pressure on one or both ovaries is well known. In the case of spontaneous somnambulism to which I have referred, I adopted her own suggestion for the purpose of relieving her attacks. A kind word, a gentle and kindly admonition frequently acts as a suggestion.

Dr. Lièbault, of Nancy, in a study on zoo-magnetism, states that M. Longpretz, a mesmerist, of Liège, who had assisted at his séance, was firmly convinced that power was transmitted from one person to another. This conviction was produced by his being able to cure infants through applying one hand to the painful or diseased part and the other on the opposite side of the body during ten minutes twice a day for a few days. Dr. Lièbault was reminded by this incident of having, many years previously, observed similar experiments by Dupotet. He tried the experiment himself and succeeded in 45 cases, of which 30 were under three years and 30 of this age and over.

Dr. Lièbault's observations are especially worthy of consideration as he has devoted many years to the study of Mesmerism and Braidism, and like many others has denied the transmissibility of power by touch, his experiments have, however, convinced him that both suggestion and touch are curative agencies.

Surgical operations have been repeatedly performed painlessly on persons in an artificial sleep. Lafontaine

---

* These remarks on Fascination are from Bottey, who copied them from Bremaud's paper, *Des différentes Phases d'hypnotisme et en particulier de l'etat de Fascination.—Bull. de la Soc. Historique*, 1884.

records two cases of parturition gone through painlessly in the hypnotic state.

Braid advocated the use of hypnotism in a variety of nervous disorders such as neuralgia, chorea, sleeplessness, &c. And his views have been confirmed by many physicians.

In the mental diseases of melancholics, monomaniacs, and others, suggestions produce changes by substituting for erroneous fixed ideas such as are rational.

Even in cases where a fixed erroneous idea has taken possession of a weak mind, repeated assurances to the contrary have been the means of an effectual cure. A case of paralysis, induced by fear of not being able to walk, in a person whose brain was weakened by work and worry, cured in this way, is recorded by Russell Reynolds.

My brother, the late Dr. Roth, of Paris, in his treatise *L'Histoire de la Folie Musculaire*, which gained the prize of the French Academy of Medicine, has described the various forms of epidemic chorea among persons making pilgrimages to religious shrines. Some of the victims kept turning round, others running to and fro and making every variety of abnormal movement.

Many of the marvellous cures reported to have occurred at Lourdes and similar shrines are due to excitement, blind faith and the journey ; while these as well as the so-called " faith cures " are influenced by suggestion just in the same manner as suggestions operate in the hypnotic state.

Bread pills and coloured water are but suggestions, and when given to hypochondriacs are as efficient as when suggestions of a more direct kind are made to an ordinary patient who is hypnotised.

Cures in some cases can be effected by inducing, during hypnotism, a hallucination that the patient is another person.

In the wards of Professor Bernheim of Nancy was a hysterical woman, devoid of sensation on one side and subject to convulsions. She was capricious and obstinate, refused her soup, which she declared to be nasty, and which she was sure she would vomit. Ordinary influences having failed, the professor placed her in a state of somnambulism and suggested that she should eat. This failed, and he resolved to change her personality. " What is your name ? " he said ; " Mary M." was the reply. " You are mistaken, you are not Mary

M., you are Josephine Durand, the aunt of Mary M." he rejoined. She allowed that she was the aunt; then said the professor " Now give your niece, who is here, a lesson, show her how to eat her meal, tell her that she is to be good and to do as you wish." The patient, in her hallucination, said her niece spoke to her, scolded her, then encouraged her to eat her food, and showed her how to do so by taking it herself and saying " That it was excellent." On awakening the patient could not believe that she had taken and retained the soup. Several similar experiments have been made with the same patient and always successfully.

Dr. Voisin of the Salpêtrière cured a case of hallucination and mania by making the patient fall into a prolonged deep hypnotic sleep, and suggesting the time for awakening. The same physician has, by making suggestions during hypnotism, altered the character. He thinks that great moral good may be accomplished in this way; and the same idea was broached by Dr. Philipps (Durand de Gros) in 1855.

Repugnance to the use of tobacco and alcohol has been induced by suggestions made during hypnotism. This leads me to notice the importance of the medico-legal aspect of hypnotism. Suggestions to commit a crime might be made during the hypnotic state, and the crime be committed practically in a state of unconsciousness. This is a matter worthy of most serious consideration.

M. Berillon has shown, through hypnotism, the functional independence of the two hemispheres of the brain, proving that the psychical, motor and sensory activity of one hemisphere could be suppressed, or that a different degree of activity could be given to each.

Bottey was present when Dr. Luys placed a woman named Valentine in a cataleptic state, and the automatism of imitation was shown on one side, while on the other an unconscious condition was induced. Approaching the left ear with the question "What is your name?" she repeated immediately with a short voice, " What is your name?" But the question being put on the right side, she replied at once, " Valentine." A scratch on one side of the face made her look angry—made on the opposite side, she laughed and looked merry.

M. Beaunis has observed that a post-hypnotic suggestion has been carried out 172 days after it was made. The same observer has modified the rhythm of the heart by suggestion, diminishing its pulsations from 98 to 92, and then increasing them to 111 and 115.

Professors Bourru and Burrot have produced epistaxis at a given time by suggestion (made during somnambulism) to the effect that it would then occur.

Stigmata on the skin have been occasioned in the same way (Bottey, from *Progrès Médicale*, 1885).

Professor Bernheim (*Progrès Médicale*, 1885) reports the cure of three cases of writer's cramp by suggestion during hypnotism.

Dr. Mabille has relieved and cured protracted sleeplessness by suggesting sleep.

At the Medico-Physiological Society of Paris, last November, Dr. A. Voisin, of the Salpêtrière, related the details of three cases of amenorrhœa, of from three to six months' duration. Each patient was placed in a state of somnambulism, and during this condition it was suggested to them that the period should commence at a certain hour, and in each it did so punctually. In two instances neuralgia was associated with the uterine fault, and this too was cured.

Dr. Voisin said that he restricted himself simply to detail the facts, leaving the explanation to future physiological enquiry, adding that previous observations by Liébault, Beaunis, Focachon, Dumontpallier, Bourru, Bourrot and Mabille had proved the possible influence of hypnotic suggestions on the vaso-motor system.

These various instances of the therapeutic use of suggestions, made during the hypnotic state, will, I trust, be sufficient to excite your interest in this subject, and induce you to study some of the many works regarding it which have lately appeared. By so doing you will be able to cure many neuro-pathic and other disorders which you probably now find difficulty in dealing with successfully.

Many cures effected by mesmerisers owe their successful result to various modes of inducing artificial sleep and by post hypnotic suggestions ; certain nervous attacks accompanied by the so-called aura, the hysterical globe in the throat, or by the hysterical nail on the top of the head were cured merely by the affirmation and suggestion in persuading the patient that there will not be another attack.

Gesner (the priest) used to exorcise people by reading and quoting to them certain parts from the Bible.

Suggestive therapeutics cure even at the present time many nervopathic complaints, while bread pills and coloured water are believed to have produced excellent curative effects in hypochondriac patients and such as suffer from imaginary disorders, it is desirable rather to confirm them in their morbid notions than to try to counteract them. If they are hypnotised their individuality might be changed and post-hypnotic suggestions made which will induce them to change their fixed ideas or to act in a manner differently from the promptings of their imaginary state.

In the service of Dr. Luys in the Salpêtrière was a girl of sixteen years of age who could not be hypnotised. She suffered from a complete hysterical paralysis with anæsthesia and had been confined to bed for fifteen months; the muscles of the legs were considerably atrophied; neither the *bromide of potassium*, douches, static and faradaic electricity, nor any other treatment caused improvement.

On the 23rd January a draught was ordered which was called *fulminante* (fulminating) and consisted of common water coloured with some harmless substance, the patient was told that it was a very powerful poison.

On the 24th, in the morning, she complained of having passed a completely sleepless night, her face was much congested, she was tormented by constant palpitation of the heart and her left arm was quite paralysed for a few hours, but without anæsthesia. The result of the imaginary poison was more violent than was expected; it was desirable that the moral causes destined to influence the brain should calm down like medicinal influences. The patient was left quiet for several days in order that she might be well prepared for the experiment.

On the 30th her diet was reduced, and on the 1st February three pills (bread) of the size of a pin's head were ordered, and she was told that they contained some of the strongest poison which had ever been known; the doctors affirming at the same time that she would be very ill, but that it was certain she would be cured; the patient was so very much afraid that it required some force to make her swallow the pills. Two hours after having taken them she had a kind of syncope, her skin got very pale, her heart palpitated, there was constant vomiting with very violent colic; these symptoms lasted almost twelve hours; afterwards creeping and pricking sensations in the arms and in the paralysed legs were felt and the next day some movements could be observed in the legs. During the following few days the power of movement increased to such an extent that she was able to walk, first with the assistance of two persons and, later, on crutches; finally on the fifteenth day, movements and sensation were perfectly recovered in the previously paralysed limbs; she pretended that even at that time the little pills still continued to act on her nerves.

Many similar cures are known where influences acting upon the imagination have caused such a revulsion in imaginary and neuropathic diseases; it is well known that Boerhave, in the case of schoolgirls suffering from chorea, caused by imitation, stopped the further progress of the epidemic by threatening to apply the actual cautery.

In a pamphlet on " Fear and Fright," published many years ago, I have collected numerous cases of complaints produced by the effect of the mental influence caused by fear and fright, and also many cases in which they had a curative effect. If suggestions, affirmations and threats can produce such effects in persons who cannot be and are not hypnotised, it will be easily understood and admitted how much more powerful the influence of suggestions will act therapeutically on those in the hypnotised state.

Dr. Voisin, of the Salpêtrière, was able to cure by the help of the hypnotic sleep a patient suffering from hallucination and a furious frenzy. He caused her to fall into a prolonged sleep and suggested the exact hour for awakening. Dr. Voisin has also modified the character of some patients mentally affected by the suggestive method; thus some who were very rude became equally polite, others who were idle began to work with much zeal. It was enough to suggest to them, while in a state of somnambulism, the idea of working at dressmaking as soon as they awoke. After a few séances of hypnotisation accompanied with suitable suggestions, the good habits continued and the whole being of the patients became absolutely changed. Voisin believes that there is reason for hoping that post-hypnotic suggestions will be able to suppress or diminish evil instincts.

Dr. Liébault, who made use of these suggestions, was able by hypnotising a lazy child to render it industrious to such an extent that he became the first in his class.

These fits of industry lasted for some time, when the child fell back into his lazy state and afterwards objected to be again hypnotised.

I have not mentioned the importance of the medico-legal aspect of hypnotism, but it is easily understood that there might occur numerous cases in which persons previously hypnotised will be irresistibly induced to commit any misdemeanour or crime which has been suggested to them. Such persons will have no recollection of the actions they have committed, and they cannot be made responsible for any committed in a state of unconsciousness. Therefore it is desirable that the attention of the legal profession, as well as of judges and juries should be directed to the importance of this subject.*

In catalepsy, phenomena consequent on different suggestions will be produced in one half of the body ; in somnambulism illusions of different kinds will occur in different places.

Dr. Magnin has frequently relieved contractions following hystero-epileptic attacks by the aid of hypnotism. Dr. Dumontpaillier has caused supplementary hæmorrhage by suggestion. From a therapeutic point of view it is remarkable that similar hæmorrhages can be stopped by suggestion. Hitherto the various explanations of the effects of hypnotism on the nervous centres which have been proposed have not been considered satisfactory, and it is still left to future experimenters to find out the real scientific explanation.

Mental Disease treated by Dr. Voisin.

" The first lunatic I have treated in this manner was one named Jeanne Sch. . . aged 22 years, who was attacked by maniacal paroxysms which succeeded hysterical attacks, hallucinations of hearing and absolute incoherence in her speech and actions.

One day she was seized by a most violent agitation for three hours. When I came to the hospital I had heard her cries at least 300 mètres off. I found her in a straight jacket, held sitting on a sofa. I tried to hypnotise her, and I succeeded, notwithstanding the difficulty of making her fix her attention on my forefinger held

---

* Gilles de la Tonresse, *L'Hypnotisme et les Etats Analogues au point de vue medico-legal.* Paris, 1886.

above the bridge of the nose. She fell into a deep sleep, and quiet succeeded immediately to the most violent agitation any one could imagine.

The first sleep lasted three hours and a half. I repeated the treatment, and I succeeded every day, but not without difficulty, for it was necessary to take sometimes two or three hours to induce sleep.

This treatment was continued without interruption; sleep lasted ten to twelve hours at least, in every 24, by means of the method of suggestion, that is to say, by enjoining the sleeper to wake at such an hour.

I am deeply astonished to be able to state that the patient gradually calmed, and that the hallucinations diminished, then ceased.

Several times I came without being expected, and I have found the patient in a paroxysm as furious and frightful as can be imagined, and hypnotism has caused it immediately to subside. It is a most touching sight to see the patient fall into a most calm sleep, and into absolute insensibility, who a minute previously gesticulated, struck and vociferated. This power of hypnotism impressed me to a high degree.

But Schaff. . . remained unsubdued as soon as she awoke, and her language and bearing were then very bad.

The idea occurred to me to suggest to her during her hypnotic sleep, ideas of obedience, of submission and good behaviour with the nurses and with us, and to enjoin her not to speak again in language improper and low, not to get into a passion, and to execute such or such work at a given time. My injunctions, as well as those of M. Gomet, my house surgeon, have been punctually executed, and I succeeded so far as to make her sew for an hour or two of a day in the work-room of my quiet patients.

She often answered us during her sleep that she would not obey, but on insisting, we obtained her promise of executing the suggestion, and she did so with astonishing punctuality and precision.

She has become obedient and submissive to rule. She does not use improper language, she keeps her person clean, and even with a certain degree of niceness.

Having obtained such results, I suggested that she should learn some passages from a moral book and recite them to me two or three days after at an hour indicated. She has done it, and has shown in reciting these passages a retentive memory more remarkable because the pages were composed of a series of detached sentences, and this girl had not read a line for several years. I have suggested to her to come and recite some chapters of morals before my pupils. She has executed my injunction.

I thought after this I would try and revive her sentiments of affection, which were absolutely extinct. She had spoken to me with hatred of her sisters. She had threatened to kill them, and had refused to see them. I have enjoined her during one of her sleeps to write me a letter, in which she will promise me to behave as a honest girl just like her sisters, and to welcome them. She has written the letter at the hour named, and on the following morning she had received her sisters with affection. Her demeanour towards them has not altered since.

So as Braid has observed, my patient had no knowledge of what passed in her hypnotic sleep, thus she cannot tell me why she did such and such acts, nor why she recited to me whole pages; she said she did it of her own accord, having no consciousness that she had done it by suggestion.

This woman was totally cured, and what is more, her whole bearing became so good that she could be admitted as attendant in one of the Paris hospitals, where her conduct is irreproachable.

Hypnotism in this case has been a means of curing madness and a moralising agent."

HYSTERIO-EPILEPSY, CONVULSIVE ATTACKS, HALLUCINATIONS OF SIGHT AND HEARING. FURIOUS DELIRIUM, MANIACAL AGITATIONS.

" A woman, G—, 25 years old, had a grandmother, on her mother's side, who was epileptic. She began to be ill five years ago when she had convulsive attacks, which were soon complicated by hallucination and delirium.

When first hypnotised she had fearful hallucinations of sight and hearing, and the delirium was as furious as possible ; she spat into the doctor's face, tried to bite and injure him. She said that " the man who is up there does not permit me to be put to sleep." She said "I am intoxicated, he does not wish me to drink or eat. I am ill-treated, and one of the students after having passed the night with me wheeled me about and put my bed into the water ; during the night I have seen serpents of various colours creep on my bed, but specially one red serpent." She told the doctor that he was a thief, a murderer, and accompanied these words with menaces; her looks were very wild.

This patient has, since she entered the hospital, attacks of mania lasting 8 to 15 hours, She was hypnotised during her period in November 1884. Afterwards this was continued after her periods as a preventive means.

The first trials were very difficult and most fatiguing. The patient was kept by five or six nurses while the doctor tried to fix her eyes on a magnesium lamp or on his own eyes, and was obliged to keep her eyelids open either with his fingers or with an instrument, and many seances have lasted one hour or one hour and-a-half and even three hours, especially in the beginning. The transition from agitation to sleep has always been very rapid, and as soon as she was asleep her voice had a sweet sound and amiable tone.

The insensibility and her collapse was complete; afterwards it was suggested that she should sleep till the next morning till 8 o'clock, that is during twenty-three hours, and at that time either the lady superintendent or the doctor would place a hand on her forehead.

If no convulsion occurred, the suggestion succeeded, but it often happened that an attack awoke the patient. Then a student replaced her into sleep. She awakes as after a natural sleep, opens her eyes, sits up; the face is animated and coloured, sometimes it is even purple, which contrasts very much with the discoloured or pale yellow tint which she had, like all hypnotised persons, during sleep.

During the hypnotic sleep, and according to the suggestion given, she takes her meals and medicines which are brought to her, the same which she has refused while awake. She leaves her bed to satisfy her bodily wants.

During the periods of agitation the patient is not permitted to be awake more than half an hour in twenty-four hours. During this half-hour she remains usually very calm, but if a longer time passes she begins again to be agitated and to be a maniac.

After a certain period of treatment the agitation has ceased, although some attacks returned; it is necessary to watch that the agitation should not reappear, and to prevent the maniacal habitude to establish itself. Thus during seven days the patient has several times remained asleep except during seven half hours.

The patient eats the hospital diet and takes *bromide of potassium* during the sleep which she refuses when awake. The refusal of eating and taking medicines is still due to the hallucination of the man who is high up, who does not permit her to take nourishment.

While the maniacal state was decreasing she was only eighteen hours hypnotised in the twenty-four. During this period she is now also quiet, but if she is permitted during the maniacal period to be more than half an hour awake, the agitation returns with the accompanying bad behaviour and words.

After four months treatment the patient had no further maniacal periods, and at present fifteen months have passed since her cure, and the attacks have ceased. The patient is now polite, sociable, and even amiable. She is grateful for my attendance, and has neither hallucination nor delirious conceptions, and she no longer believes me to be the assassin or the murderer of her father. She is employed in the Salpêtrière as linen-room keeper" (Voisin).

THERAPEUTIC APPLICATION OF HYPNOTISM (AZAM).

It is impossible to know beforehand to what extent the study of Braidism will contribute to relieve and cure many complaints. Braid quotes 65 observations of cures of various diseases. The thoughtful medical man will no doubt ascribe much to the enthusiasm of the inventor of hypnotism.

A method which causes at our will anæsthesia hyperæsthesia, the most absolute immovability of one or another part of the body, which retards or increases as required the circulation, which can produce a calm sleep and arrests, as M. Peul and Azam have seen, spontaneous catalepsy, &c. Such a method must have a certain therapeutic future as soon as it is studied without prejudice and merely with the desire to find out the truth.

An important question must be asked, Is there any danger in hypnotism? The abuse of its application might fatigue the nervous system and cause hysterical attacks. It is not prudent to employ it in epileptic patients nor on those who suffer from heart disease. ? Neither Braid or Azam have ever observed that life is endangered by hypnotism, and the latter has never observed even fainting. Only medical men should make use of it as they alone can judge whether there are any counter indications, and will be able to relieve nervous accidents. The whole future of surgical anæsthesia produced by hypnotism depends on patient and well-made experiments; the operations which have been already made prove sufficiently that insensibility to pain can be realised.

Surgical Application of Hypnotism.

It is only necessary to substitute for the sponge or the compress impregnated by *ether* or *chloroform*, a very bright, shiny object, for the purpose of putting a patient to sleep. The principal condition is to fix the attention of the person, to have the mind calm, there must be no noise and therefore it is desirable that the patient should be several times hypnotised before any operation is performed, in order to be sure that he can be put to sleep and that the period of the anæsthesia is sufficiently long. But surgeons in general object and find that *chloroform* is much easier to use, but as we practise surgery not for ourselves but for the sake of our patients, it is desirable to have some perseverance to avoid any movement in the shining object and to prevent any noise, because some patients are sometimes quickly excited by anything which affects their hearing. A medical man who was hypnotised by M. Verneul, and who is able to give a good account of the first period before he fell asleep, affirms that the slightest movement of the shiny object kept before his eyes or even a very slight noise retarded and prevented him from falling asleep.

Medical Precepts regarding the Application of Suggestion. By Prof. Bernheim.

" Shall we prescribe something which might be useful although its abuse might be injurious ? Thus should we not prescribe wine, alcohol, opium, quinine, because the immoderate use and unsuitable use of these substances might produce accidents ; there is no doubt that suggestion applied by dishonest and unskilful people is a dangerous practice.

The law can and must interfere to prevent abuse.

Suggestion applied for curative purposes and used with prudence and intelligence can only be beneficial.

It is the task of the medical man to apply its useful effect for the relief of his patients. If I think that suggestive therapeutics has some chance of success I believe I deserve some reproach if, as a medical man, I would not propose it to my patient, and if I would not insist upon inducing him to try it.

But before I apply it, I consider it my duty to adhere to some rules which every medical man should strictly follow to preserve his conscience and his professional honour.

1.—Never to put to sleep any person without his formal consent or the consent of those who have the charge of the patient.

2.—Never to cause sleep except in the presence of a third authorised person, relative, husband, father, &c., who guarantees both the hypnotiser and the hypnotised. Thus every disagreeable supposition or future accusation will be prevented, as well as every suspicion of any experiment the aim of which is not for the good of the patient.

3.—Never to make to the hypnotised person without his consent any other suggestions than those required for his cure.

The professional man has no other powers than those which are confided to him by the patient. He must limit himself to merely therapeutic suggestion, not make, even if it is in the interest of science, any experiment except after previous formal consent of the patient.

Further, the medical man must not make use of his influence upon the patient for obtaining this consent, when he thinks that the experiment which he intends to make might be in any way inconvenient.

As a physician and clinical professor it was my duty specially to study suggestion applied for curative purposes.

I have the right to affirm, supported by numerous facts, that curative suggestion exists, without wishing to say that it is always applicable or always efficacious. But this is often the case. It is not because I am indolent nor is it that I wish merely to satisfy a vain superficial scientific curiosity that I began this study about five years ago, which I have pursued vigorously, notwithstanding many obstacles and notwithstanding the ridicule to which I was exposed " (Bernheim).

APPENDIX I.

Cures by Miracles and Imagination.*

The stones attached to the girdles of the Hebrew pontiffs and of the priests of Cybele, the stones under the form of a beetle, of a hand, of a circle which were worn round the neck of Orientals, of the Greeks and Romans, and even the cameos worn at the present time by ladies have been considered by the Ancients as magic talismans of mysterious power.

Paracelsus, who has observed the cause of the effects produced by amulets and similar objects, says: "Whether the object of your belief is real or false, you will always obtain the same effect; thus, if I believe in a statue of St. Peter, just as if I believe in St. Peter himself, I shall obtain the same effects as if I would have believed in St. Peter himself; but this is a superstition. It is the belief which produces these miracles and whether the belief is true or false it will always produce the same wonders.

Peter Pomponazzi, an author of the 16th Century, says:—

"We can easily understand the marvellous effects produced by confidence and imagination, especially when they are mutual between the patient and the person who acts upon them. The cures attributed to certain relics are the effects of this imagination and of this confidence. The wicked and philosophers know that if you place the bones of any skeleton in the place of the bones of a saint, the patients will still recover their health as long as they believe themselves to be near the real relics of a saint.

Cures by Mineral Magnetism.

The loadstone which was used by the Egyptians in the fabrication of their prophylactic amulets has in all times cured gouty pains, headache, toothache and hysterical complaints.

During the last century Father Hell, the well-known Viennese astronomer, has prepared artificial magnets which cured spasms, convulsions and paralytic affections.

The magnetic contrivances, constantly and habitually applied by the Abbé Lenoble have also been very efficacious in severe nervous affections.

* From Prof. Bernheim.

The treatment by magnets and by metals according to Dr. Burcq have been the real precursors of the hypnotic treatment. Mineral magnetism preceded Mesmer's animal magnetism, and the modern magneto-therapeutics preceded the curative suggestion.

In 1777 the Paris Royal Medical Society examined the effect of Lenoble's more perfect artificial magnets which had such a power as had never been before obtained.

Drs. Audry and Thuret have been commissioned by the Society to repeat Lenoble's experiments, and after having made forty-eight observations they confirmed the statement that these magnets relieved or cured toothache, nervous headache, rheumatic pains, facial neuralgia, tic-douloureax, stomach pains, convulsive hiccup, palpitations, different kinds of trembling, convulsions, hystero-epilepsy, etc., etc.

Amongst the observed cases a large number were relieved shortly after the application of the magnet; strong neuralgia and facial pains have been relieved each time by contact with the magnet.

Spasmodic and convulsive symptoms disappeared rapidly after the magnet has been applied, a nervous cough was calmed instantly and did not occur again.

One case of convulsive movements of the arm and of its contractions which prevented the use of the hand was relieved considerably in the course of the day. Rheumatic pains were soothed, and if they returned after the displacement of the magnetic contrivances they disappeared again as soon as the magnet was replaced. In toothache the relief was sometimes immediate, sometimes the magnet did not produce any relief while similar pains in other patients ceased by the contact with the magnet; it sufficed to prolong the application of the magnet or to apply a stronger one when relief was brought on.

Sometimes the magnet increased the pains or appeared to produce sensations which had not been felt before.

In one case the magnetic band caused fever and headaches which ceased on the removal of the apparatus. An epileptic patient felt slight and continual fainting which ceased when the magnet was removed, and the epileptic fits appeared to be more intense.

Another patient suffering from nervous paralysis, suffered from similar faintings from the use of the magnet.

Various sensations have been observed when the contrivances were changed or their application interrupted, as for instance, sensation of heat in the diseased parts, giddiness, nausea, palpitation, headaches, itching, etc.

The Commissioners of the Royal Medical Society declared that these mineral magnets had a real effective power in soothing or curing nervous affections of various kinds, but they did not relieve organic diseases.

Many other observers have confirmed the truth of Audry and Thuret's observations. Trousseau mentioned in 1833 that he has sometimes made use of the magnet, and that he can affirm that it has a curative effect on the parts with which it is in contact, and that it has an influence which it is impossible to ascribe merely to the imagination of the patient; he has seen neuralgic pains relieved and fits of nervous dyspepsia suddenly arrested.

Larnce praises the effect of magnets in angina of the chest, and Trousseau confirms the soothing effects in various maladies; it is incontestable that temporary cures have been made in rheumatism and the case is mentioned of a French marshal whose rheumatic pains could not be relieved in any way except by the application of magnets. Notwithstanding Trousseau's assertions these magnetic cures have been almost entirely forgotten, and a similar disbelief in animal magnetism and Mesmer's manipulation has also prevailed for a long time.

CURES BY METALS. (BURCQISM.)

When Dr. Burcq published his doctrine of metallotherapie he was also disbelieved, and he had to wait till the physicians in the Salpêtrière began again to make experiments with his treatment by metals, and then only the belief in the curative effect both of metals and magnets was revived.

I myself had occasion during Dr. Burcq's visit to London, about thirty years ago, to assist at his experiments in the mesmeric hospital, which was established at that time, especially by Drs. Elliotson and Ashbourne.

Dr. Elliotson has published at that time and at my suggestion a long report on metallo-therapeutics in the Zoist.

I have also made use of the various metals in form of bracelets, armlets and bands, which have been applied to the thighs, calves and arms.

For the purpose of finding out which metal would reproduce sensation in an anæsthetic part, small metallic stripes of about an inch in width and three or four inches in length, have been placed on each finger of a hand which had lost sensation, on each finger a different metal was applied, and after about ten minutes the little metallic bands were removed and it was found that under the influence of one of these metals, the sensation had returned; if of the five different metals none was æsthesiogen, other five metals were chosen in order to find out which would be the suitable one in the given case.

At that time Burcq had no indication for choosing the suitable metal except by applying the various metals experimentally. Long needles of gold, silver and steel have been used for the purposes of testing the sensibility of the patient. As there was usually a diminution of muscular power connected with anæsthesia, Burcq had a small dynamometer for testing the degree of pressure produced by the grasp of the hand. The small metallic bands, the needles, and the dynamometer were neatly placed together in a little box 3 inches square, which could be carried in the pocket.

As soon as it was ascertained which metal produced the sensation, it was chosen for the larger metallic application on the forearm, upper arm, and the lower extremities; patients suffering from neuropathic complaints, loss of sensation and of power in various parts, were soon relieved if the suitable metal was found, but frequently the pains returned after the removal of the metal. As I have not pursued my studies and observations on this subject, I am unable to speak of the present state of metallo-therapeutics, which during the last few years has been used successfully in cases of anæsthesia, but since Dr. Liébault has made known his suggestive method of treating nervous complaints, suggestions have been used instead of mineral magnetism and metals.

The invocations of Egyptian priests for the purpose of obtaining the cure of limbs submitted to the influence of the various deities using the magic formula, teaching the use of simple remedies; the sympathic powder of Paracelsus, Perkins' metallic tractors, and the pseudo metallic, that is, wooden tractors of Drs. Hoggarth and Falconer, and at present the so-called electro homœopathic medicines of Count Mattei, the king's touch, the celebrated cures made at the tomb of the diacre Pares and those of Knock in Ireland, as well as those at Lourdes in France, are proofs of cures performed by imagination or suggestion.

Many healers and curers, believed themselves possessed of supernatural powers, have no doubt cured many people without knowing it by mere suggestion; thus the Irishman, Greatrakes, the German priest, Gesner, the Prince Abbot of Hohenlohe, Father Matthew, a peasant in the neighbourhood of Saumur, the Zouave Jacob, are quoted by Bernheim amongst those who have cured by suggestion. Dr. Hack-Tuke mentions Dr. John Tanner as an enthusiastic advocate of the treatment of hysterical aphonia by electro-magnetism applied only to the tongue; among fifty cases not one was unsuccessful, and he mentions that in one case the return of the voice showed itself by a great cry, in another case the voice returned immediately, in a third, although the voice returned it disappeared again, but returned again after a second application of the electro-magnetism, in a fourth case there was also an instantaneous return of the voice; but Dr. Tanner insists on the great importance persuading the patient that electro-magnetism will cure him, and it is probable that if he is not persuaded that there will be a failure.

The treatment by suggestion is not new, it is its methodical application and its definitive adoption in the ordinary practice which is new.

Dr. S. Obernheim mentions a medical man who attended a patient suffering from paralysis of the tongue who had been unsuccessfully treated by various methods. The medical man had invented an instrument from which he expected excellent results, and he intended to try it, but wishing to know the patient's temperature he placed a thermometer into the mouth, the patient fancied that this is the wonderful instrument which would cure him, and in the course of a few minutes, he called out with great joy that he can freely move his tongue.

A similar case is related by Bernheim, under whose care a young girl was placed in the hospital who, during the last four weeks, suffered from a complete nervous aphonia, after the professor had made his diagnosis, he told the students that this disease is sometimes instantaneously cured by electricity, but which can act through his suggestion. The instrument was sent for, but before making use of it the professor wished to try simply a suggestion by affirmation; he placed his hand on the larynx, and while moving this organ, he said " Now you can speak very loudly." In an instant he ordered her to say A B, afterwards Mary, and thus she continued to speak distinctly.

The aphonia had disappeared.

Another effect of imagination during sleep on the peristaltic action of the intestines is mentioned in " La Bibliothèque chosie de Médicine."

The daughter of the Consul of Hanover, 18 years old, was ordered to take an aperient, viz., rhubarb, towards which she had the greatest antipathy.

She dreams that she has taken the abhorred medicine. Influenced by this imaginary rhubarb she awoke, and the bowels acted several times.

A similar result is mentioned by Demangeon, a monk ordered to take medicine the next morning, dreamt that he had done so, and when he awoke the dream acted as if he had really taken the medicine.

Amongst the moral causes appealing to the imagination and influencing the cerebral mechanism of possible cure, none is so efficacious as religious belief, to which are due many well authenticated cases.

The Princess of Schwartzenberg was affected for eight years by paraplegia, the most celebrated German and French doctors have been consulted; in 1821, Prince Hohenlohe, who was a priest, brought a peasant to the princess, who had convinced the young priest of the power of prayer in the treatment of diseases. The Princess was stripped of all mechanical apparatus which Dr. Heine had applied a few months before, to prevent the contraction of the limbs. The priest asked the patient to join her belief to his and to that of the peasant and to pray. After a short time the Princess was addressed in this way. "Do you believe that you are relieved?" "Oh yes, I believe it most sincerely!" Then the priest said, "Rise now and walk." At these words the Princess rose, made a few turns in the room, tried to go up and down stairs. Next day she went to church and from that moment she retained the use of her limbs (related by Charpignon).

APPENDIX II.

HYPNOSCOPE.

This is an instrument invented by Dr. Ochorovicz who asserts that all persons who are influenced by the magnet can be hypnotised in proportion to their sensitiveness to the magnet. The hypnoscope is a tubular magnet of three or four centimetres in diameter and five to six centimetres in length, in which an opening is left of two centimetres. The weight of the whole is about 169 grammes.

Steel of Allevard is used for its construction, and the instrument has the power of supporting 25 times its own weight.

The following is the way it is used; after the armature (closing piece) is removed the index (first finger) is placed inside the tube so that it touches both poles the finger being with the palmar surface upwards, and the instrument hanging on it; after two minutes the finger is taken out, and the modifications produced in it are examined.

Ochorovicz, who has been making experiments since 1880, has never observed any new sensation, and he does not believe that a magnetic sense exists, but this does not prevent 25 per cent. of men being influenced by an average powerful magnet, who feel certain very distinct sensations.

Twenty in 100 persons have sensations of formication and disagreeable pricking, sometimes like sparks and needles pricking the skin. Seventeen in 100 have a sensation of a cold breath or sensation of warmth and dryness. These two impressions can exist together, the one in the right arm the other in the left arm. It is known that a magnet placed under the sole of the foot of some paralysed persons can warm the patients, while a good fire had not such an effect.

The sensation of the cold breath is very much like that felt when near an electro static machine.

Eight in 100 have painful sensations, pains in the joints. They say that they had a feeling of broken bones.

Five in 100 have sensations of swelling of the skin, swelling which may be real, because it is sometimes difficult to withdraw the finger of the subject from between the poles of the hypnoscope.

Two in 100 have sensations of heaviness in the finger or the whole arm.

A person otherwise in good health was asleep, the eyes shut, the pupils turned upwards, the head completely enveloped in a thick veil, at each approach of the magnet, a distance of 15 centimetres, the hand was drawn quickly towards himself, and followed all his movements until the moment when it became rigid and insensible. Then it was necessary to restore sensibility or rather hyperæsthesia to recommence the experiment, I must add that the same phenomenon was reproduced although perhaps more feebly, by a metal, a glass, or some other body being placed in contact. The person questioned in his sleep said that he had felt attracted in a given direction without knowing why.

Symptoms Produced by the Hypnoscope.

The objective modifications are more important for the diagnosis. They belong to one of the four following categories.

a. Involuntary movement (very rare).
b. Insensibility (analgesia or anæsthesia complete).
c. Paralysis (impossibility to move the finger).
d. Contraction (rigidity of the muscles).

The symptoms disappeared after a few minutes under the influence of a slight massage, and without this they can last several minutes and even several hours.

The persons to whom the hypnoscope causes insensibility, paralysis or contraction, can be hypnotised at the first sitting, in others the experiment must be repeated.

There is no doubt that the magnet has a certain influence on the nervous system of persons who are predisposed.

But whether this magnetic action is really incontestible, this can be only partially admitted because

1. The importance of the influence is not in direct proportion to the power of the magnet, but rather in proportion to the degree of the hypnotic sensibility of the person on whom the experiment is made.

2. The north pole has not a different influence from the south pole, which would not be the case if it were a purely magnetic effect. Is it perhaps a metaloscopic influence as Mr. Pellat has suspected? To this question, which was asked by Dr. Brown-Sequard, on the day

when Dr. Ochorovicz made his first communication, the answer is

1. That the number of persons who were sensible to metals is smaller than of those who are influenced by the magnet, and still smaller than of persons who are influenced specially by steel.

2. There are persons influenced by metals, for instance by copper, but not influenced by the magnet, and therefore not influenced by hypnotism.

But is this mysterious action perhaps only imaginary or suggestive as it is at present called?

This is not the case. (1) Because it sometimes though rarely influenced persons without their knowledge, also persons who are asleep and also animals.

It is easy to distinguish the sensations produced merely by an emotion or by an " expectant attention " from those which are produced by the hypnoscope itself; the imaginary sensations change their character or disappear at a new trial, while the real sensations repeat themselves, preserve their character, and get more and more distinct.

The imaginary sensibility disappears, while the real sensibility increases by force of habit.

Gessmann, in Vienna, has invented a Hypnoscope, in which four magnets are arranged in such a manner as to leave in the centre a space through which the index is introduced, and thus is placed in contact with all the poles of the four magnets.

TABLE OF CONTENTS.

ANALYTICAL INDEX.

WORKS BY Dr. ROTH.

Prize Medals have been awarded at the International Exhibitions of London 1862, and Paris 1867, for Dr. Roth's Works on Physical Education, Models, &c., also a Gold Medal at the Health Exhibition 1884.

ON THE MOVEMENT CURE.

1. A SHORT SKETCH OF RATIONAL MEDICAL GYMNASTICS, OR, THE MOVEMENT CURE. With Thirty-eight Engravings. 1s.

2. THE PREVENTION AND CURE OF MANY CHRONIC DISEASES BY MOVEMENTS. With Ninety Engravings. 10s.

3. THE HANDBOOK OF THE MOVEMENT CURE. With One Hundred and Fifty-five Original Engravings. 10s.

3B. BRAND'S TREATMENT OF FEMALE DISEASES BY THE MOVEMENT CURE. Translated by Dr. Roth. 5s.

ON PARALYSIS.

4. CONTRIBUTION TO THE HYGIENIC TREATMENT OF PARALYSIS, AND OF PARALYTIC DEFORMITIES. With Thirty-eight Engravings. Illustrated by Numerous Cases. 8s. 6d.

5. ON PARALYSIS IN INFANCY, CHILDHOOD, AND YOUTH, AND ON THE PREVENTION AND TREATMENT OF PARALYTIC DEFORMITIES. With Forty-five Engravings. 8s. 6d.

6. Contributions to the Treatment of FACIAL PARALYSIS. 6d.

6B. On WRITER'S CRAMP, with Engravings. 1s.

ON SPINAL CURVATURES.

7. THE PREVENTION AND TREATMENT OF SPINAL DEFORMITIES, ESPECIALLY OF LATERAL CURVATURE, with notes on the causes, the artificial production, and injurious modes of treatment of these complaints. With Fifty-three Engravings. 5s.

8. Notes on Sayer's TREATMENT OF ANGULAR CURVATURE. 6d.

9. ON LATERAL CURVATURE. NOTES ON THE EXAMINATION, PROGNOSIS AND TREATMENT. 1s.

10. ON THE INFLUENCE AND USE OF THE WILL IN THE TREATMENT OF MANY SPINAL DEFORMITIES. 6d.

BALNEOLOGICAL PAMPHLETS.

11. MEDICAL AND OTHER NOTES, collected on a Holiday Tour to Arcachon, Biarritz, Pau, and the principal Watering Places in the Pyrenees. 1s. 6d.

12. DIVONNE-LES-BAINS IN SPRING AND AUTUMN. Notes on its Sanatorium and Hydropathic Establishment. 6d.

13. THE HOT SPRINGS OF BOURBONNE-LES-BAINS. (The French Kreuznach.) 6d.

14. THE RUSSIAN BATH, published with a view to recommend its introduction into England for Hygienic as well as curative purposes, with some suggestions regarding Public Health. Second Edition. 1s.

MISCELLANEOUS PAPERS.

15. ON HAHNEMANN'S MERITS, ERRORS, AND CRITICS. 1s.

16. ON FEAR AND FRIGHT AND THE DISEASES THEY CAUSE AND CURE. 1s.

17. BONE SETTERS AND RUBBERS. 6d.

18. ON COMPULSORY MEDICAL INSPECTION ON ALL SCHOOLS. Paper read at the Société Médicale du Bureau de Bienfaisance, 13 Octobre, 1880. 6d.

19. ON THE PREVENTION OF BLINDNESS. 6d.

20. THE FEVER-DENS IN THE NORTH-WESTERN AND WEST CENTRAL DISTRICTS IN LONDON. 6d.

ON SCIENTIFIC PHYSICAL EDUCATION AND HYGIENE.

21. ON THE IMPORTANCE OF RATIONAL GYMNASTICS AS A BRANCH OF NATIONAL EDUCATION : a Letter to Lord Granville. 1s.

22. ON SCIENTIFIC PHYSICAL TRAINING AND RATIONAL GYMNASTICS : a Lecture delivered at the request of the Committee, at the Royal United Service Institution. 1s.

23. A PLEA FOR THE COMPULSORY TEACHING OF PHYSICAL EDUCATION. 1s.

24. ON THE LADIES' SANITARY ASSOCIATION (of London) AND THEIR WORK. Paper read at the International Congress of Hygiene in Turin in 1880.

25. ON THE NEGLECT OF PHYSICAL EDUCATION AND HYGIENE as a Cause of the Deterioration of the Physique of the Population. 1s. 6d. and 2s.

26. ON SCIENTIFIC PHYSICAL EDUCATION AND ITS PRACTICAL INTRODUCTION INTO SCHOOLS. An Address delivered at the request of the Committee before the Conference of Teachers, January, 1880. 6d.

27. L'EDUCATION PHYSIQUE SCIENTIFIQUE ET L'HYGIENE SCOLAIRE. Lettre à M. le Ministre de l'Instruction Publique en France, 1880. 6d.

28. L'INTRODUCTION DES ELEMENS DE L'HYGIENE ET DE L'EDUCATION PHYSIQUE SCIENTIFIQUE DANS TOUTES LES ECOLES PRIMAIRES ET SECONDAIRES. Paper read at the International Congress of Hygiene in Turin, in 1880. 6d.

29. GYMNASTIC EXERCISES WITHOUT APPARATUS ACCORDING TO LING'S SYSTEM. With Forty-two Illustrations. For the Development and Strengthening of the Human Body in Childhood and in Youth. Sixth Edition. 1s.

30. A TABLE OF A FEW GYMNASTIC EXERCISES WITHOUT APPARATUS. With Thirty-three Wood Engravings. 6d.

81. **Table**, showing a few Injurious Positions to be avoided durir time of Education and growth, and some deformities of tl Spine, produced partly by bad positions, and partly l tight lacing. With Forty-six Engravings. 6d.

82. **Two Tables of Gymnastic Exercises without Apparatus;** wit Explanations. Mounted on Cloth, with Rollers, Varnishei Per set 8s. 6d.

83. **The Gymnastic Exercises of Ling**, arranged by Rothstein translated by Dr. Roth. Second Edition. 2s. 6d.

84. **Gymnastic Exercises on Apparatus**, according to the Rationi System of Ling, with 80 Illustrations. 1s.

85. **Gymnastic Games.** These consist of a series of Thirty-six Cardi Each Card presents an accurate representation of som position in which the body is to be placed, as a Gymnasti Exercise. The Cards are accompanied by instruction explaining the methods of playing various amusing Games and at the same time they are most efficient in promotin a healthy development of the muscles, and form a syster of physical training. In a cloth case 1s. 6d.

86. **Gymnastic Models**, copied from Dr. Roth's Collection in tl South Kensington Museum. These Models are executed i: Papier Maché, each twelve inches in height. The price c the set of Twenty-four, mounted on hard wood, is £3. i selected set of Twelve Models on a slightly smaller scal may be had, price 30s.

87. **On the Causes of the Great Mortality of Children and th Means of Diminishing them.** A paper read before th Scientific Congress at Rouen, and the International Medica Congress in Florence, in 1869. Also a French, Italian and Hungarian translation. 6d.

LONDON:

BAILLIERE, TINDALL & COX, 20, King William Street, Strand
A. N. MYERS & Co., 15, Berners Street, Oxford Street, W.